My
Little Golden Book
About
BUGS

By Bonnie Bader
Illustrated by Emma Jayne

The editors would like to thank Dr. David Grimaldi, Curator,
Division of Invertebrate Zoology, at the American Museum of Natural History,
for his assistance in the preparation of this book.

A GOLDEN BOOK • NEW YORK

Text copyright © 2020 by Penguin Random House LLC.
Cover art and interior illustrations copyright © 2020 by Emma Jayne.
All rights reserved. Published in the United States by Golden Books, an imprint of
Random House Children's Books, a division of Penguin Random House LLC, 1745 Broadway,
New York, NY 10019. Golden Books, A Golden Book, A Little Golden Book, the G colophon,
and the distinctive gold spine are registered trademarks of Penguin Random House LLC.
rhcbooks.com
Educators and librarians, for a variety of teaching tools, visit us at RHTeachersLibrarians.com
Library of Congress Control Number: 2019933464
ISBN 978-0-593-12388-1 (trade) — ISBN 978-0-593-12389-8 (ebook)
Printed in the United States of America
10 9 8 7 6 5 4 3 2 1

They creep and crawl. They fly and flutter. They can come in beautiful colors. Some stink and some sting. What are they?

BUGS!

Let's go find some.

Look! Down on the ground—it's a superhero! What? Where? That little ant?

Yes! An ant may be tiny, but it is very strong. An ant can carry fifty times its body weight. That's like a three-year-old child carrying a cow!

Ants live in groups called colonies. Some colonies can be underground or in a tree. The queen ant's job is to lay eggs. The worker ants look for food, clean the colony, and take care of the babies. Some worker ants are called soldiers. They protect the colony. If there is danger, they can bite or sting an attacker.

air defense system

main entrance

entry point

entrance

egg hatchery

royalty room

living space

There are more than 12,000 kinds of ants in the world. And there are more than 350,000 kinds of beetles. That's a lot of beetles!

Most beetles have a hard shell and two sets of wings. They can be many colors, including yellow, green, orange, red, and purple. The front wings form part of the shell, and the back wings make them fly.

Some beetles, such as Colorado potato beetles, are harmful to plants. These beetles eat potatoes and other crops, such as eggplants, peppers, and tomatoes.

But other beetles are helpful. Ladybugs eat bugs such as aphids that can hurt plants in your garden. And young dung beetles eat poop. Without dung beetles—and dung flies—there would be a lot more poop on the earth!

Did you know that a firefly is a type of beetle? Fireflies are also called lightning bugs.

A male firefly will light up when he wants a female to see him.

Stink bugs look like beetles because their front wings are hard. But they have different ways of eating. Beetles have hard jaws for chewing. Stink bugs have a beak for sucking up liquid.

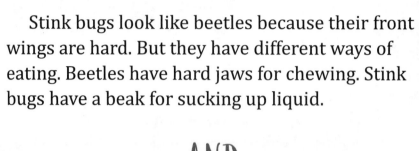

AND
STINK BUGS STINK!

If a stink bug thinks it is in danger, it releases a super-nasty scent.

When a stink bug isn't making a big stink, it's busy eating plants, leaves, and flowers. These little bugs are less than an inch long, but with their big appetites, they can cause a lot of damage to farms and gardens.

CREEP, CREEP, CREEP.

A caterpillar creeps along on three pairs of legs. As it creeps, it eats, eats, eats. And as it eats, it grows, grows, grows.

When the caterpillar stops growing, it turns into a pupa. And when the pupa pops open, out comes a butterfly or a moth!

At first, the butterfly's or moth's
wings are soft, wet, and wrinkly.
Once the wings expand and dry,
the creature takes to the sky.

This bug may look like a butterfly, but it's an atlas moth.

The atlas moth is one of the biggest bugs in the world. Its wingspan measures ten inches!

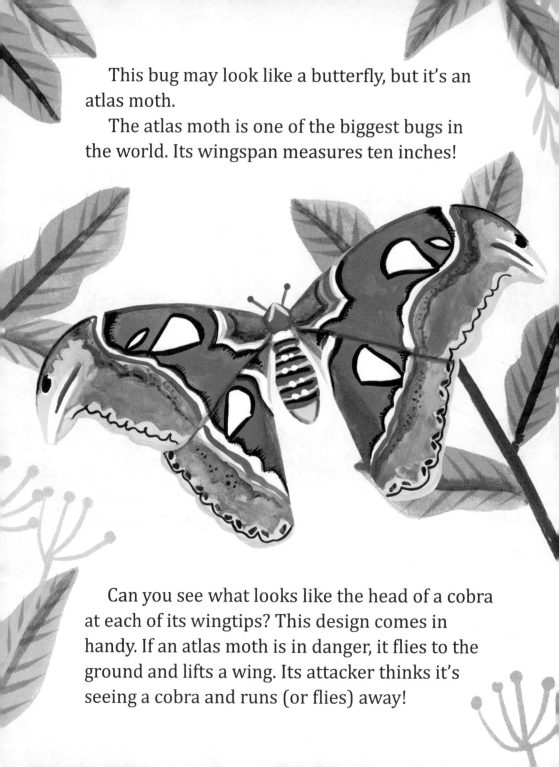

Can you see what looks like the head of a cobra at each of its wingtips? This design comes in handy. If an atlas moth is in danger, it flies to the ground and lifts a wing. Its attacker thinks it's seeing a cobra and runs (or flies) away!

What has six legs but can't walk very well?
A dragonfly!

A dragonfly is a much better flier than a walker. In fact, it is one of the fastest flying bugs in the world! And it eats as it flies, snagging small insects right out of the air.

Dragonflies have see-through wings, but the rest of their bodies are very colorful. They can be blue, green, yellow, or red.

When this bug bends its front legs, it looks like it's praying. That's why it's called a praying mantis.

A praying mantis has very large eyes. This is helpful when it looks for food. Crickets, grasshoppers, and flies are common meals for a praying mantis. Really big mantises can even catch small lizards and birds!

When it's mealtime, the mantis grabs
the prey with its front legs. The spikes on
its legs pin it down—so the praying mantis
can bite off the head first!

And praying mantises like to stay clean.
They groom themselves, like cats do.

Which kind of "bug" eats flies? A spider!

All spiders have eight legs and are covered with little hairs.

Some spiders make silk and use it to spin sticky webs. When a fly or another type of bug lands on the web, it gets stuck—and becomes the spider's meal.

The largest spider in the world is
a kind of tarantula.

Would a spider or another bug or animal like to eat a stick? Probably not. But look closely. This is not a stick. It is a stick insect! This bug blends into its surroundings so animals won't eat it.

Other bugs are also good at hiding to protect themselves. Can you spot a false-leaf katydid on a twig? Do you see any lantern flies on the tree trunk? Is there an orchid mantis hiding on a flower?

This black-and-yellow bug is easy to spot.

A honeybee can fly for miles every day, sipping nectar from flowers. Once the bee is full, it flies back to its hive, where the nectar is used to make honey, which feeds the young bees.

We've seen tiny bugs and giant bugs;
bugs that hide, and bugs that light up.
Some are helpful, while others cause harm.

WHEREVER YOU LOOK,
BUGS ARE ALL AROUND!